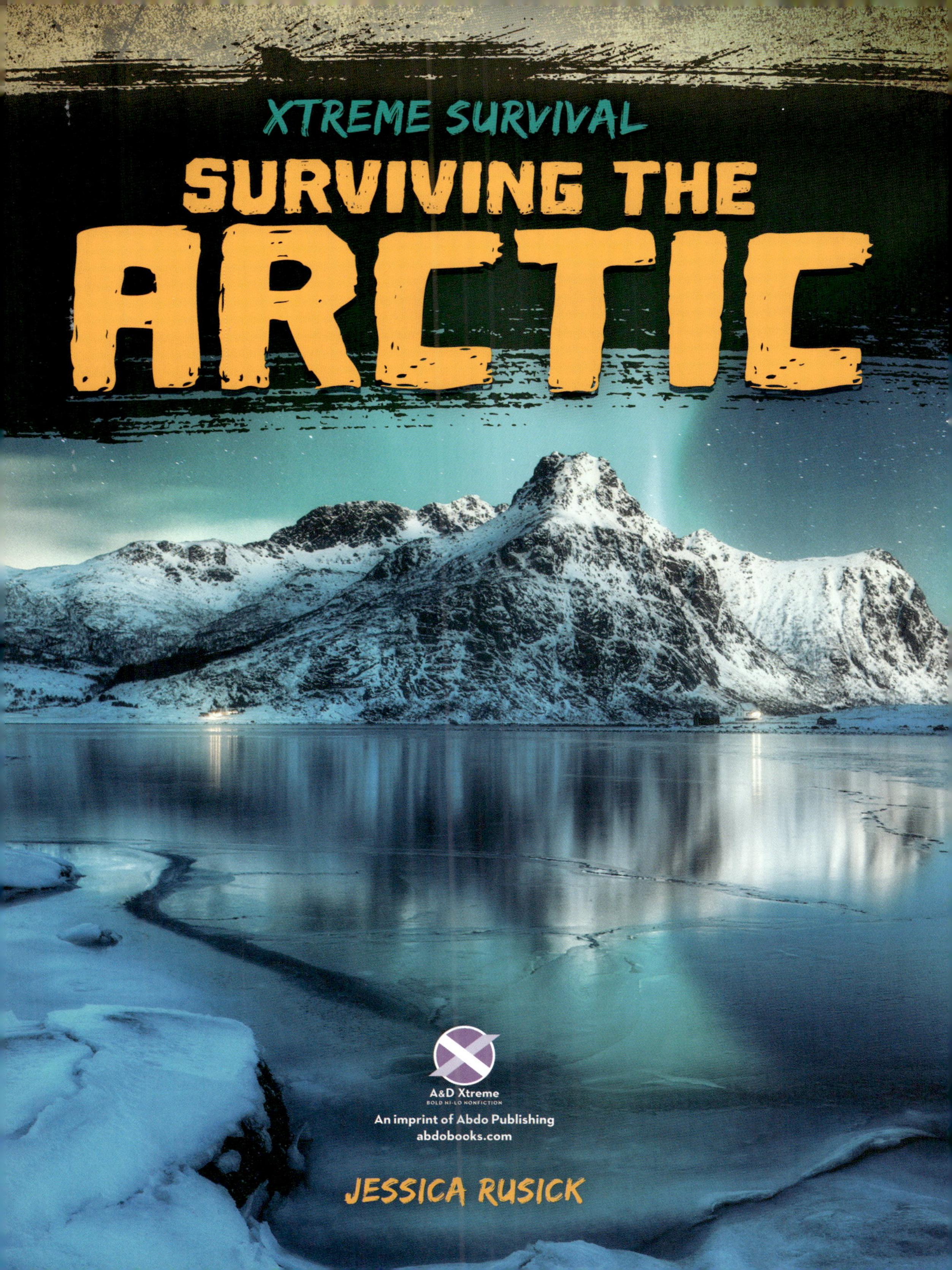

XTREME SURVIVAL
SURVIVING THE ARCTIC
A&D Xtreme
BOLD HI-LO NONFICTION
An imprint of Abdo Publishing
abdobooks.com
JESSICA RUSICK

TAKE IT TO THE XTREME!

GET READY FOR AN EXTREME ADVENTURE!
THE PAGES OF THIS BOOK WILL TAKE YOU INTO
THE THRILLING WORLD OF OUTDOOR SURVIVAL.
WHEN YOU HAVE FINISHED READING THIS BOOK, TAKE THE
XTREME CHALLENGE ON PAGE 45 ABOUT WHAT YOU'VE LEARNED!

ABDOBOOKS.COM

Published by Abdo Publishing, a division of ABDO, PO Box 398166, Minneapolis, Minnesota 55439.
Copyright ©2024 by Abdo Consulting Group, Inc. International copyrights reserved in all countries.
No part of this book may be reproduced in any form without written permission from the publisher.
A&D Xtreme™ is a trademark and logo of Abdo Publishing.
Printed in the United States of America, North Mankato, MN.
102023
012024

THIS BOOK CONTAINS RECYCLED MATERIALS

Design: Tamara JM Peterson, Mighty Media, Inc.

Editor: Katherine Chu

Cover Photograph: Denis Belitsky/Shutterstock Images

Interior Photographs: art of line/Shutterstock Images, pp. 12–13; Bain News Service, publisher/Wikimedia Commons, p. 7; borchee/iStockphoto, pp. 28–29; Darwel/iStockphoto, p. 44; Denis Belitsky/Shutterstock Images, p. 1; EB Adventure Photography/Shutterstock Images, pp. 42–43, 46; FedBul/iStockphoto, pp. 38–39; ginger_polina_bublik/Shutterstock Images, pp. 8–9; isabel kendzior/Shutterstock Images, pp. 32–33; Joy Prescott/Shutterstock Images, pp. 4–5; Julian Idrobo/Wikimedia Commons, p. 34; Maridav/Shutterstock Images, pp. 24–25; Megan Frost Photography/Shutterstock Images, pp. 20–21; mmac72/iStockphoto, pp. 34–35; muratart/Shutterstock Images, pp. 40–41; Ninok_stock/Shutterstock Images, pp. 36–37; photos_martYmage/iStockphoto, pp. 18–19; ra-photos/iStockphoto, pp. 26–27; Smith Archive/Alamy Photo, p. 5; sodar99/iStockphoto, pp. 14–15; Terje Lein-Mathisen/iStockphoto, pp. 16–17; Timaldo/Shutterstock Images, pp. 22–23; Vera Vakulova/Shutterstock Images, pp. 6–7; wanderluster/iStockphoto, pp. 30–31; Wikimedia Commons, pp. 10–11

Design Elements: Nik Merkulov/Shutterstock Images (grunge background; queezz/Shutterstock Images (brush texture)

Library of Congress Control Number: 2023939333

Publisher's Cataloging-in-Publication Data

Names: Rusick, Jessica, author.

Title: Surviving the arctic / by Jessica Rusick

Description: Minneapolis, Minnesota : Abdo Publishing, 2024 | Series: Xtreme survival | Includes online resources and index.

Identifiers: ISBN 9781098291815 (lib. bdg.) | ISBN 9781098278717 (ebook)

Subjects: LCSH: Survival--Juvenile literature. | Survival skills--Juvenile literature. | Arctic Regions--Juvenile literature. | Cold adaptation--Juvenile literature. | Outdoor life--Juvenile literature. | Wilderness survival--Juvenile literature.

Classification: DDC 613.69--dc23

TABLE OF CONTENTS

CHAPTER 1

AN ARCTIC EXPEDITION

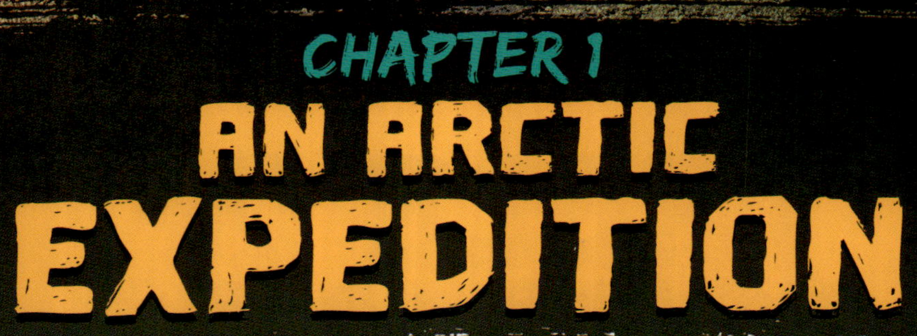

In 1921, Ada Blackjack joined four men on an expedition to Wrangel Island in the Arctic Ocean. The 23-year-old Iñupiaq woman was hired to cook, sew, and perform other household tasks. The four men were largely inexperienced and unprepared for the journey. Blackjack was unsure about going, but she needed the money to care for her son.

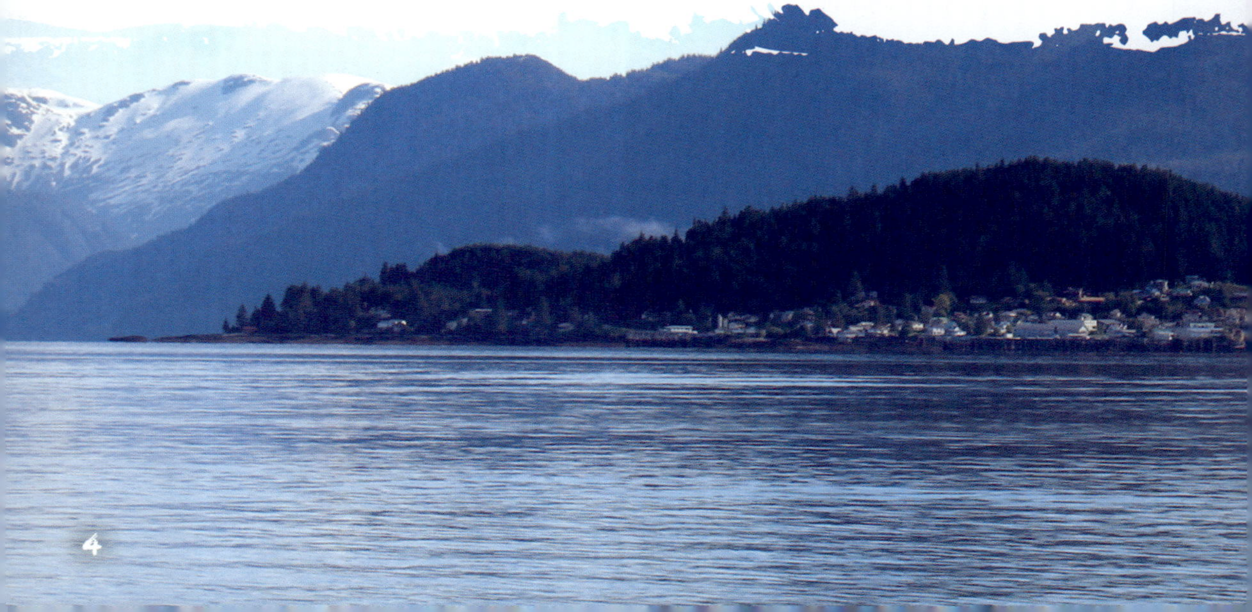

Blackjack (*left*) left her son at an orphanage until she was able to make enough money to take care of him.

In September 1921, the group arrived on Wrangel Island. They set up camp, hunted, and explored. The winter was rough, supplies **dwindled**, and the expected resupply ship could not reach them due to heavy ice. By January 1923, the group had begun to starve. Three of the men left to find help while Blackjack cared for the fourth man, Lorne Knight, who was ill. The men never returned, leaving Blackjack and Knight **stranded** in the Arctic. How would they survive?

Vilhjalmur Stefansson, an Arctic explorer and anthropologist, organized and helped fund the expedition to Wrangel Island.

XTREME FACT

Blackjack had no experience surviving in the wilderness before going to Wrangel Island.

The Arctic is a **dangerous** place to become **stranded**. Freezing temperatures take a **toll** on the body. They can cause hypothermia, or low body temperature. There is little food to eat. Hypothermia, frostbite, and polar bears are constant **threats**. But many people have overcome these dangers and more, thanks to their survival skills.

Arctic summers are cold and short. They last for three months. Usually, the temperature does not reach more than 50 degrees Fahrenheit (10°C).

SURVIVOR SPOTLIGHT

Blackjack placed warm bags of sand on Knight's feet, protecting them from frostbite. She built a boat from driftwood, animal hide, and canvas to help her hunt. She also learned to trap foxes for food. Knight died in June, but Blackjack survived for two more months. A ship finally reached Wrangel Island in August 1923. Blackjack was rescued!

The Wrangel Island expedition team was made up of (*left to right*) Allan Crawford, Fred Maurer, Blackjack, Knight, Milton Galle, and Victoria the cat.

XTREME FACT

A cat named Victoria was brought on the expedition. She survived and was rescued along with Blackjack!

CHAPTER 2

FINDING
WATER

If possible, purify water before you drink it. Although Arctic water is cleaner than water from many other environments, it can still make you sick.

Humans can only survive for a few days without water. So it's important to find water right away. In summer, rivers and lakes are the best sources of drinking water. In 2016, reindeer herder Egor Tarasov became lost in northern Siberia, Russia. He drank from rivers, surviving for 42 days!

Rivers and lakes freeze in the winter. Luckily, snow and ice are water sources. Place snow or ice in a bag, then put it under a layer of clothes. Body heat will melt the snow or ice into water! Fire can also melt snow and ice.

Around 10 percent of the world's fresh water can be found in the Arctic's ice.

CHAPTER 3
FINDING FOOD

Being in the cold reduces the body's energy, so finding food is important. Many edible plants grow during the summer, including dandelion, Arctic raspberries, and willow. Tarasov ate mushrooms and berries to survive.

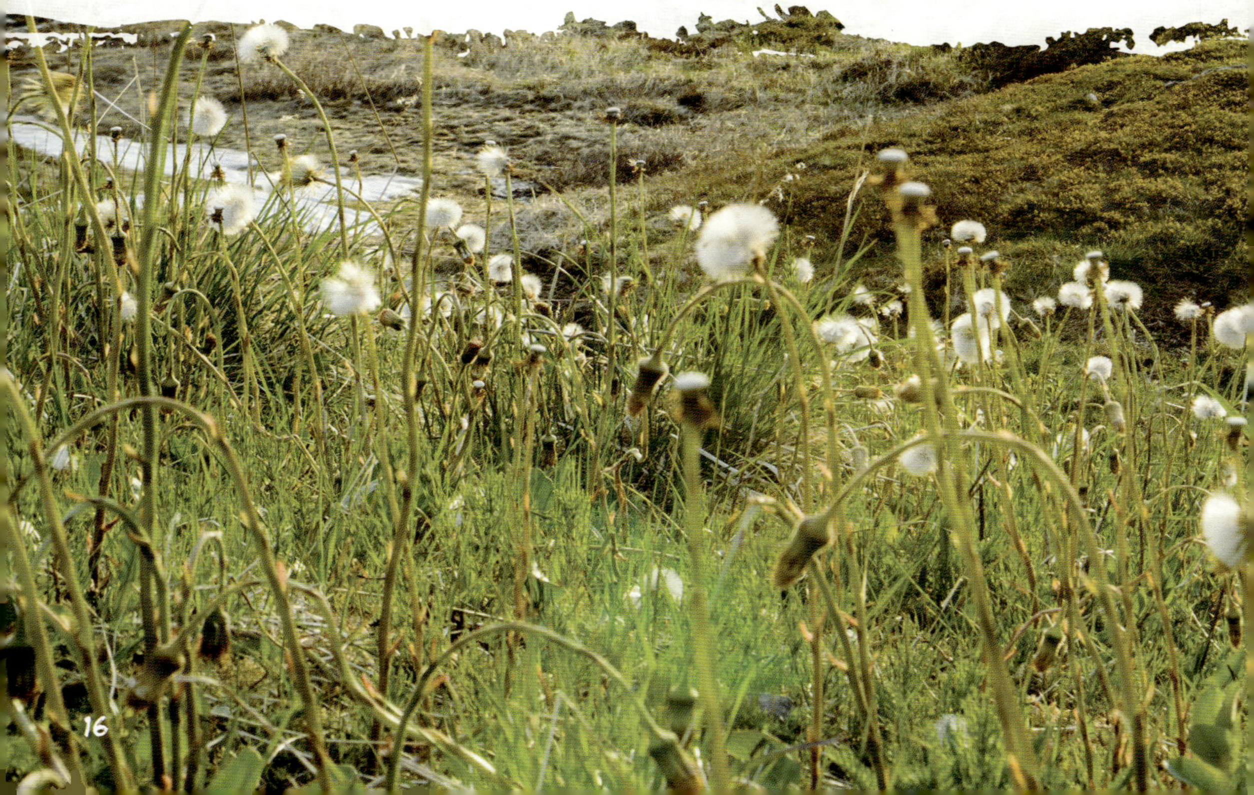

Although there are many edible
plants in the Arctic, there are
poisonous ones as well. To avoid
them, only eat plants you recognize.

Thick snow and ice cover plants in the winter, leaving fewer food options. Small animals such as hares, foxes, ptarmigan, and grouse are good food sources. Hunting these animals often requires weapons. But it's possible to trap animals with minimal supplies.

Animals in the Arctic tend to be white to blend in with the snow. So spotting them requires careful attention and patience.

One way to trap animals is with a simple wire trap. Create a small loop at one end of the wire, twisting it back around itself. Thread the opposite end of the wire through the loop, making a noose. Attach the noose to a tree. It will tighten around an animal going through it.

You are more likely
to catch an animal
if your noose is
near an animal
burrow or trail.

CHAPTER 4
STAYING
WARM

The Arctic's average winter temperature is
–30 degrees Fahrenheit (–34°C). This **extreme**
cold can cause hypothermia, which can cause
death in less than an hour.

Tarasov was stranded in Yakutia, near the East Siberian Sea. He prevented hypothermia by walking at night, when it was coldest.

Check for frostbite often.
Wiggle fingers and toes
to warm them. Place your
hands in your armpits.
Rub your ears and nose
with mittens.

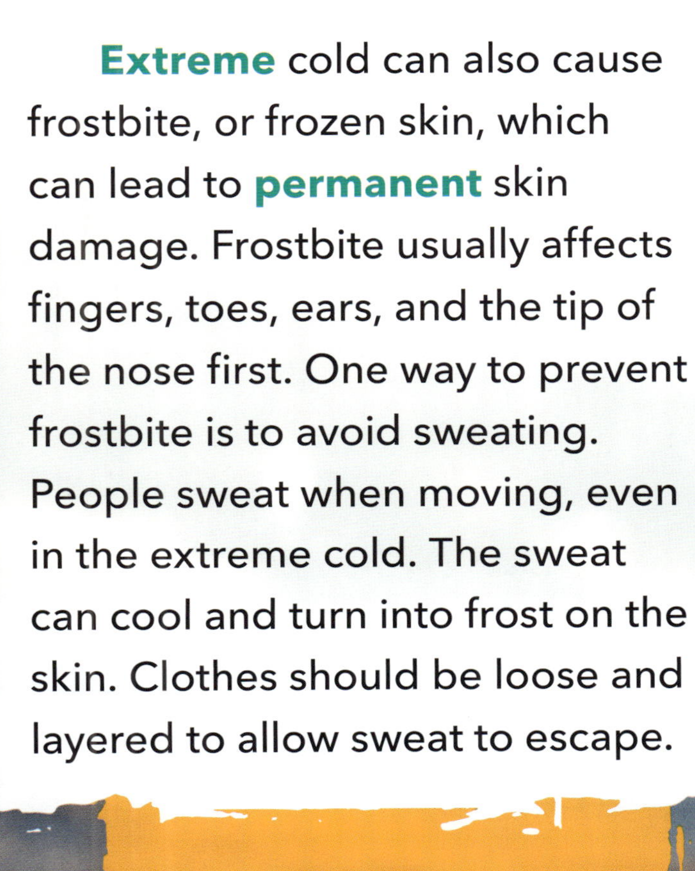

Extreme cold can also cause frostbite, or frozen skin, which can lead to **permanent** skin damage. Frostbite usually affects fingers, toes, ears, and the tip of the nose first. One way to prevent frostbite is to avoid sweating. People sweat when moving, even in the extreme cold. The sweat can cool and turn into frost on the skin. Clothes should be loose and layered to allow sweat to escape.

XTREME FACT

In extreme cases, frostbitten limbs must be amputated.

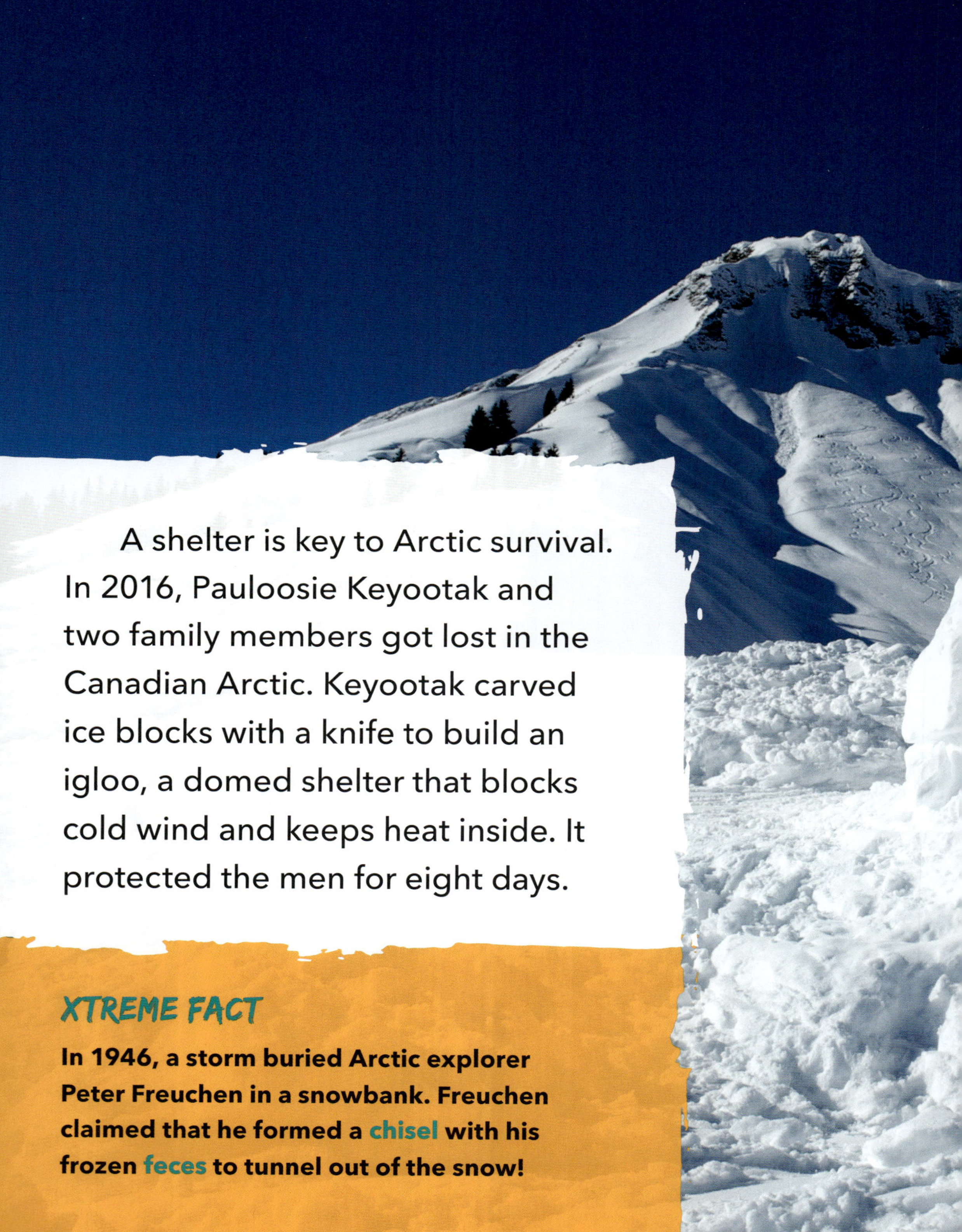

A shelter is key to Arctic survival. In 2016, Pauloosie Keyootak and two family members got lost in the Canadian Arctic. Keyootak carved ice blocks with a knife to build an igloo, a domed shelter that blocks cold wind and keeps heat inside. It protected the men for eight days.

XTREME FACT

In 1946, a storm buried Arctic explorer Peter Freuchen in a snowbank. Freuchen claimed that he formed a chisel with his frozen feces to tunnel out of the snow!

The Inuit invented igloos. They used them as temporary homes during the winter or when they hunted in colder regions.

When building a snow cave, look for
a big snowdrift. Wind can compact
snow, making snowdrift caves
stronger and easier to build.

A snow cave is a simple shelter. Dig up into a snowdrift, then create a platform for sleeping that is higher than the cave entrance. This helps warm air stay in while keeping cold air out. Or get creative! Hunter Ernie Eetak became **stranded** in a blizzard after his snowmobile broke down in 2021. He tied a tarp against his snowmobile to protect himself.

XTREME FACT

Eetak's hands were amputated due to frostbite. By 2023, he was able to hunt again with the help of prosthetics.

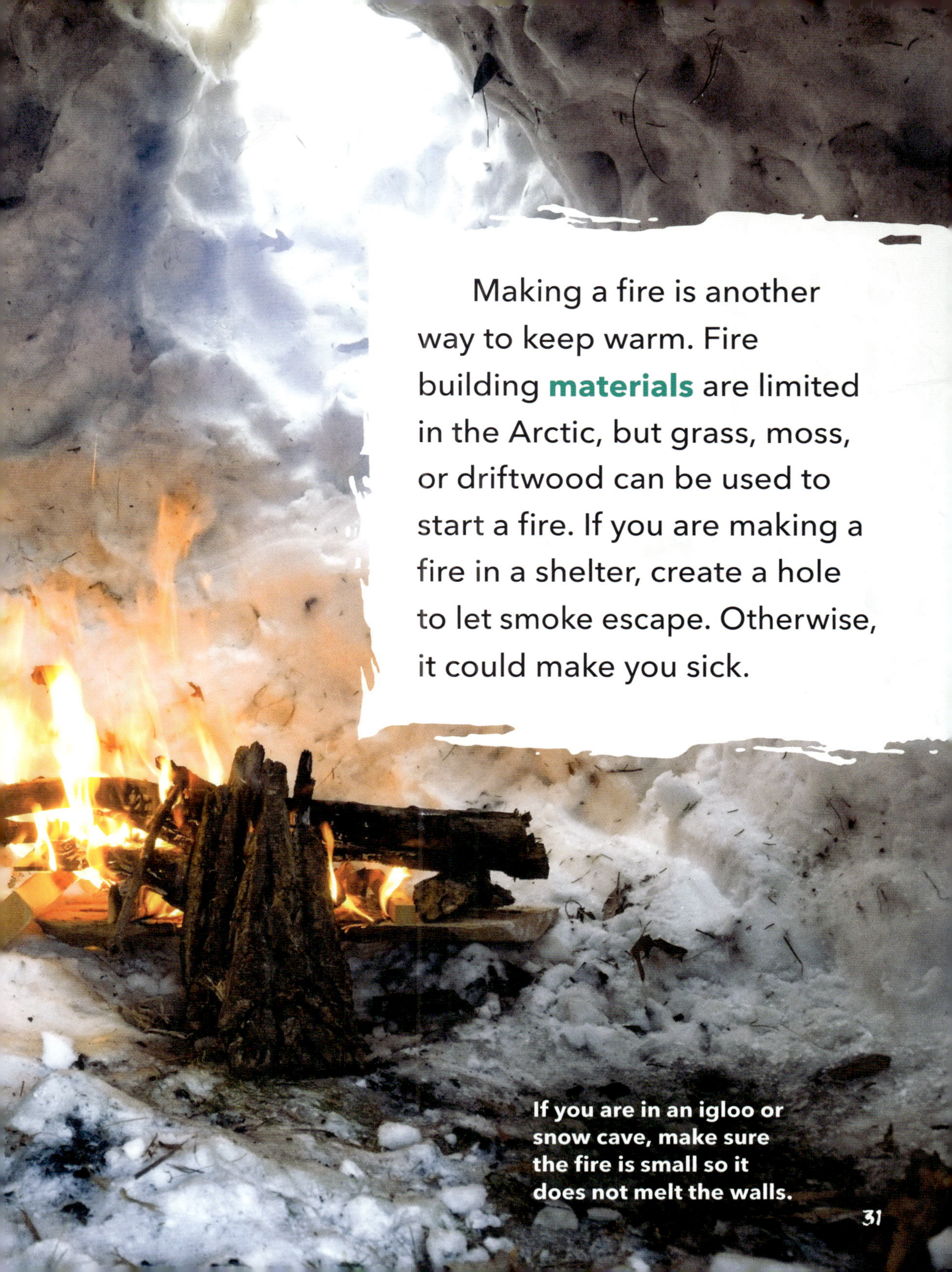

Making a fire is another way to keep warm. Fire building **materials** are limited in the Arctic, but grass, moss, or driftwood can be used to start a fire. If you are making a fire in a shelter, create a hole to let smoke escape. Otherwise, it could make you sick.

CHAPTER 5

ARCTIC DANGERS

If a polar bear attacks,
hit it in sensitive areas,
such as the eyes or nose.

Polar bears are **dangerous** predators that sometimes hunt humans. It's important to watch for polar bears, but do not approach them. Blackjack built a wooden platform as a lookout. Make loud noises to scare off polar bears that come near.

Snow goggles were originally made by the Inuit people to help them avoid snow blindness.

Another danger is snow blindness. This is temporary eye pain and loss of vision caused by the glare from white snow. Snow blindness can make it nearly impossible to **navigate**. But special goggles will help. Eetak made goggles with thin eye slits from caribou antlers to reduce snow blindness. Survivors can craft similar goggles using wood or cardboard.

Walking through deep snow will also dampen your clothes and feet, which can lead to frostbite and hypothermia.

Walking through deep snow is difficult and draining. Snowshoes help you walk on top of the snow instead of sinking into it. Gather pine branches or bundles of sticks to build snowshoes. Tie the branches or sticks to the bottoms of boots with a cord or shoelaces.

The color of ice is important. Blue
or clear ice is the thickest and
safest to walk on. Black ice is the
thinnest and most dangerous.

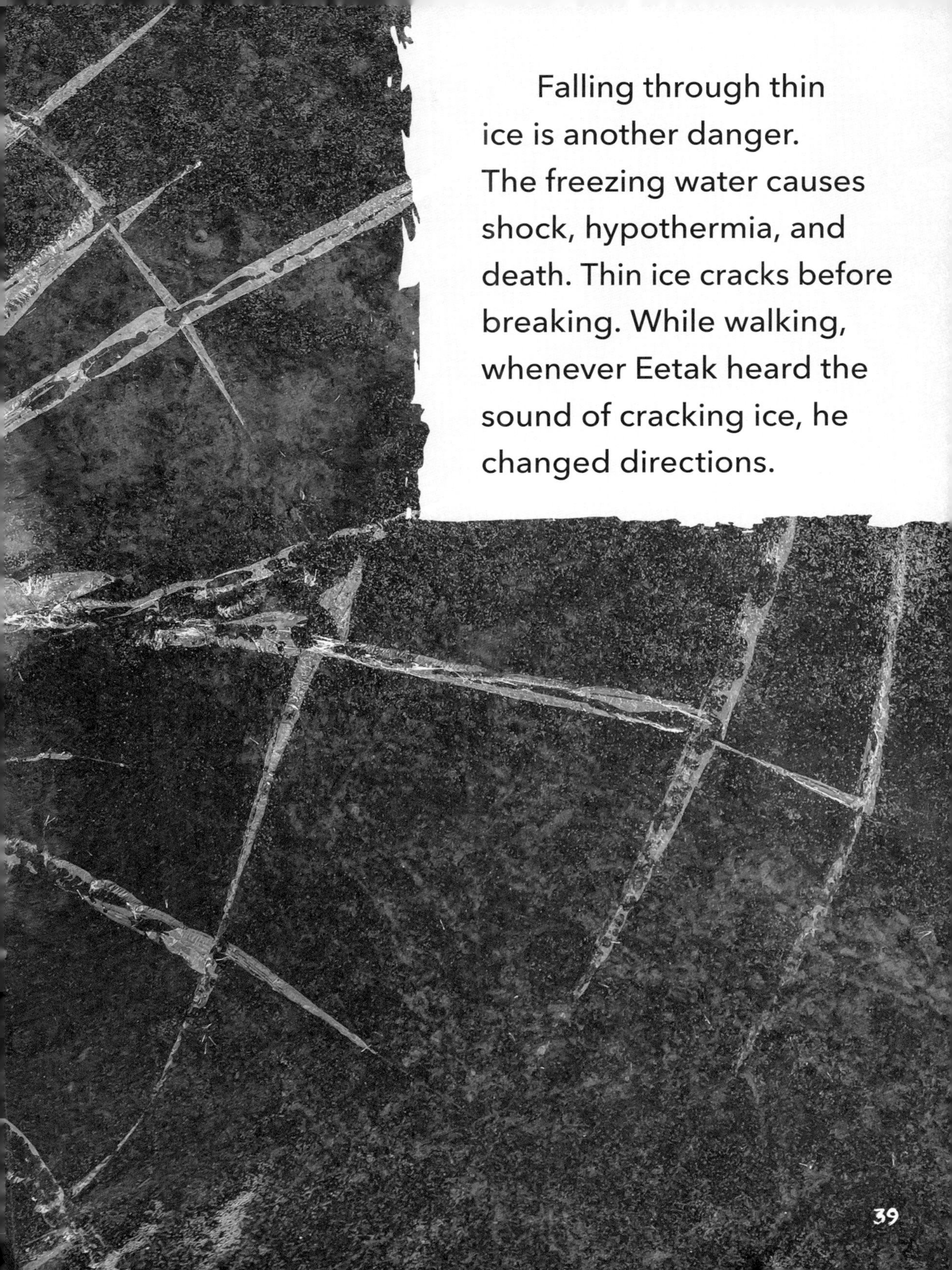

Falling through thin ice is another danger. The freezing water causes shock, hypothermia, and death. Thin ice cracks before breaking. While walking, whenever Eetak heard the sound of cracking ice, he changed directions.

CHAPTER 6

GETTING RESCUED

Getting **stranded** in the Arctic is scary. It's important to stay calm instead of giving in to fear. Thinking about his family helped Tarasov stay positive while he was stranded. In 1967, pilot Bob Gauchie's plane ran out of gas, forcing him to land in the Arctic. He was stranded for 58 days and occupied himself by writing in a diary.

A supply plane flying over the Arctic. Gauchie was rescued when a passing supply plane spotted him.

Many Arctic rescues happen by air. It's easy for rescuers to spot colors against white snow. Waving colorful items or reflecting light off a mirror can catch rescuers' attention.

The countries in the Arctic share resources for search and rescue operations.

Talon

CHAPTER 7

SURVIVING THE ARCTIC

Being **stranded** in the Arctic means having little **access** to food and shelter. **Extreme** cold is a constant danger. But with top-notch survival skills and a bit of luck, some people have overcome these challenges to survive. Think back on the Arctic survival stories you read and skills you learned. Could you survive in the Arctic?

It's good to tell at least two people where you are going and when you expect to return. That way, rescue teams will know where to look if you get stranded.

XTREME CHALLENGE

1) What is the best way to get the attention of Arctic rescuers?

2) What animal did Ada Blackjack trap when she was stranded?

3) How would you keep calm if you were stranded in the Arctic?

4) If stranded in the Arctic, would you rather have snowshoes for walking or sunglasses to protect against snow blindness? Why?

5) What areas does frostbite usually affect first?

GLOSSARY

access—the opportunity to gain or use something.

amputate—to remove, especially to cut a limb from a body.

chisel—a tool with a flat, sharp end that is used to cut and shape a solid material such as stone, wood, or metal.

dangerous—able or likely to cause hurt or harm.

dwindle—to become less.

extreme—very much, or to a very great degree.

feces—solid bodily waste.

material—what a thing is made up of.

navigate—to find the way from place to place.

permanent—lasting for a very long time.

prosthetics—the medical specialty dealing with artificial body parts, such as limbs and hearts.

stranded—lacking the means to leave a place.

threat—something that could be harmful.

toll—a cost in life or health.

ONLINE RESOURCES

To learn more about Arctic survival, please visit **abdobooklinks.com** or scan this QR code. These links are routinely monitored and updated to provide the most current information available.

INDEX